The Hitchhiker's Guide to Virtual Reality

Mfon Akpan

ST PETERSBURG PUBLISHING

First trade paperback edition July 2020

Page Design by bioGraph
Cover Design by lefthandedscissorz
Manufactured in the United States of America
1 2 3 4 5 6 7 8 9 10

Printed on 90gsm acid-free paper

ISBN: 979-8-64330182-0

To God, for His endless mercy. To my parents Jacob and Theresa. To my children Mfon and Keturah. To my dearest Vivian for your tireless encouragement and inspiration. To the rest of my family near and far, friends, and students.

Table of Contents

Preface

The purpose of this book is not to serve as a manual for implementing virtual reality. It is to give educators—K–12 and university professors—a practical and tactical method for using virtual reality in the classroom. I want educators to understand why it is important for students to be exposed to virtual reality, even if they're not able to use virtual reality in the form of headsets or other accessible technological hardware.

Students can understand, comprehend, and touch the technology with their minds and in their hearts, which will allow them to transition to using it in its ever-evolving forms. Technology is changing continuously and at a faster pace than we can ever keep up with. The current forms of virtual reality will not be the same which students will see when they get into high school or graduate from college. When we think about students as they move through the education process, it is imperative that we make sure they are exposed to technological understanding so they will be able to use technology in the future and lower the shock to the system.

The reason I say "shock to the system" is because

there may be students who have no idea that this technology exists. By being exposed to it and understanding it in the classroom, they will greatly enhance their learning journey. As educators, we owe it to them to expose them to technology and give them as many tools as possible.

What Is the Why?

When we mention the "Why," Simon Sinek comes to mind. In my case, Sinek and Dr. Kelly Richmond Pope come to mind. When I had the opportunity to give my first TEDx talk and it was accepted, I reached out to my mentor and friend Dr. Pope, a professor. I was excited with the prospect of giving a TEDx talk and called Dr. Pope to tell her I was accepted and would be giving a talk.

Here is the first thing she asked me: What is your talk about? I told her it was about incorporating virtual reality into the classroom. Here are the next questions she asked me: Why does anyone care about your topic? Why should I care about it? Why should I be interested in this? Why? Why? Why?

This was a shock to my system, and it was a wake-up call. I had to think about why anyone would care about virtual reality and why they should care about this technology in the classroom. Why is it important that people are doing what they are already doing in the classroom? Students are learning, and teachers are teaching, so why add this extra technology? If no one is using it in the mainstream classroom, why is it so important?

Her questions struck a nerve and allowed me to hone in and focus on one thing—what is the importance of this technology to students today? I needed to realize that there is a mission, and that mission is to make sure students are exposed to technology that is advancing faster than we can adapt. What we have today could be unrecognizable in two years. So it's important to expose students to technology and allow them to use it and understand it—at least being exposed to it in their minds and knowing how it works. Prior to graduating from high school and as they start college and eventually graduate, they will know what this technology is and will have seen it. They won't be starting from zero. The challenge for educators is that they have to learn the technology. That could be difficult for many because they're getting along fine without it. Adding something else, learning it, explaining it, and showing it to students could be challenging.

So here is the "Why": We owe our students and future generations exposure to this technology and the ability to use it. We have to give them the best possible start so they won't have to start at zero, even if we have limited resources at home or in the classroom.

Some schools may already have plenty of resources and students are starting at 7. Other students have limited resources and are starting at 2. And then there are other students with no resources who are starting at zero. So it's important that we learn this technology.

I am an accounting professor who has opportunities to serve students at a university in Chicago that works primarily with underprivileged, nontraditional students. To put it plainly, I work with a lot of minority students, and I also come from a nontraditional background. I have traveled to various conferences and given presentations that were part of our accounting association for professors. In 2018, I had the opportunity to present and attend the Ohio regional meeting for our association, the American Accounting Association.

At the meeting, I attended a talk by Mark Rubin, the president-elect at that time and a professor at Miami University in Ohio. This talk resonated within me and changed me, and I had an epiphany that there are students who are the haves and the have-nots. Professor Rubin was explaining how the accounting profession is changing and how, as accounting professors and educators, we need to change in order to keep our professional lives viable and be able to train the students of the future.

Dr. Rubin also mentioned that his university is creating classes for Power-BI and data analytics and pushing for these courses to train the accountants of the future. I realized how far behind the program is where I work and how far behind my students will soon be.

It really touched me in that moment that there's a gap, and I saw that there's a big difference between the students who are going to graduate from my business program and the ones Dr. Rubin was describing. There's a wide gap, a huge divide, in exposure to technology and what students are going to be able to do. Dr. Rubin was talking about one thing, and I got something completely different out of his presentation and conversation.

It made me understand that it is imperative to make sure that students have access to the latest and greatest technology. Dr. Rubin was discussing data analysis and Power BI for his undergraduate accounting students. My students have never heard about Microsoft's Power BI, let alone understand how data analytics impacts the accounting profession. Will I be able to set up a data analytics course? Will I be able to set up a Power BI course? Will I be able to set up a Tableau course at my university? That is the gulf I'm talking about, and it is huge. On the one hand, we have students graduating with bachelor's degrees in business who have taken courses in Power BI, have learned data analytics, and are learning blockchain in courses that professors have developed. On the other hand, the students I'm teaching have no idea how these things work, and they're getting the same bachelor's degrees. So who is at a disadvantage when these graduates sit down in an interview with a potential employer? There's that gulf, and it made me realize that I may not be able to give my students everything they need. I may not be able to set up a blockchain course. I may not have the resources for a Power BI, Tableau, or Abacus course or a virtual reality lab for my students. But I can do my best.

What I mean by doing my best is that I can use whatever resources I have to give the students the exposure and access they do not have. So when it comes to virtual reality, I realize I won't be able to get a budget to set up a VR lab at the school or even pay for it out of my own pocket. I realize I won't have all the means, but I do have something for my students. Maybe I don't have Oculus Rift, Oculus Quest, HTC Vive, or even Google Cardboard for each student. But I can create 360 Content and give my students some type of exposure to virtual reality in this technology.

The other thing I realized that day as I listened to Dr. Rubin is that there's a wide gulf, there's a disparity. There are the haves and the have-nots as far as access to technology and education. But here's the great thing about it and the advantage. Technology is always changing, so today we're talking about Power BI, and next year we could be talking about Abacus. This year we're talking about Oculus Go, and next year we might be talking about Oculus Quest. So technology is always changing. The great advantage is that we don't have to marry the technology but rather marry what the technology does.

It's more important to understand the key principles and the inner workings of the technology than to learn a specific program or know all the ins and outs of a specific piece of hardware. The key is to understand how in general that technology works because it's going to keep changing. There's going to be something new year after year or every other year that you're going to have to adjust to.

If you have no exposure o or understanding of the general technology, it will impact how quickly you can adapt to a new version, a new technology, or new ways things are set up. A perfect example of this is Power BI, Tableau, and a software called Abacus that is used for data analytics. The key thing, the important thing, for students to learn is coding because the types of software tools we use for managing big data will continuously change, but there are certainly key areas students should know about in order to easily adjust and adapt to the new software. If students have an understanding of coding, it will be easy for them to adapt to the new software that's used for data analysis.

Using these key principles, we can teach our students. What if you're not able to teach your students code? What's the next best thing? What if you don't have resources to teach them programming, so you expose them to this or that software so they can at least have an idea of how to use them. They can have at least a leg up. Running up a set of 50 stairs, some students have a 40-stair advantage, and some only have a 10-step advantage but are able to see what virtual reality is and how it works. The students are able to see whether it's a 360-degree video with you talking. They're able to actually see it, experience it, and understand how it works, and many of them have never even seen the technology used outside of a gaming application—maybe at Dave and Buster's—but even a 360-degree video can open them up to other applications and ways that technology can be used. I think it's important for us to implement the technology in the classroom. It's the exposure that's important.

So the Why is the future. The Why is our future generations of students we educate. Right now, I'm a university professor. What I'm teaching freshmen this year is not what I'll be teaching them in four years when they graduate. This year's information may be obsolete, and we may be using something else. In four years, I have no idea how things will be with virtual reality. This year, Oculus Quest came out, and it's revolutionizing virtual reality and opening things up for a broader audience at a higher level. Four years ago, we would not have imagined that it could even exist.

We need to find out what students have Oculus Quest or are able to create software. If they're able to have exposure to it, that's great, but we also need to integrate the technology into the coursework and give them something so it's not a shock four years down the road. What is virtual reality? Some have never seen it and don't understand it. So we have to work to give our students the advantage. That is the Why, and the Why is the future. The Why is what we owe to our students to show them what is coming, even if it's a glimpse of what is coming. They're brilliant and they're smart, and they can understand and put things together. Giving them an advantage is something we have to do. We have to make it imperative.

As educators, we need to be two steps ahead of our students. We owe our students our best efforts to literally predict the future. They need to have a whole foot in the doorway of the future. Virtual reality is currently an important technology, a disrupter and displacer of jobs, and in my opinion, it is going to be a bigger technology than artificial intelligence. It is currently being used in healthcare, and the synergies are so great for the applications of virtual reality technology that we're only scratching the surface.

We're doing a great disservice to our students by not exposing them to this technology. Our students should not get into the workforce where they're asked to train using virtual reality for their job and after four years of college have no idea what virtual reality is. That's doing them a disservice, and that is the Why. We need to make sure our students are prepared for tomorrow. They need to be exposed to virtual reality and be able to understand its fundamentals so as it develops, they will also be able to develop. And when they are in the workforce and continue their education, they will at least have some sort of foundation or rudimentary knowledge of the technology and won't be starting from ground zero and allow students from other universities who have had this exposure to get the upper hand. So when asked what the Why is, the Why is our students, and our students are our future. Virtual reality should be incorporated into the classroom because our future depends on it, and our future rests on our students and future generations.

As developers and enthusiasts of virtual reality technology, we are always looking to push technology to the highest level and farthest point. We need to realize that many of us are very early adopters of virtual reality technology, which means that from an educational standpoint, we have to realize that the technology we're working with today may be obsolete in the next two years or so. We have to focus more on the skills and exposure than the technology. The key is not the technology. The key is the skills that people need to have, which can come from exposure to the technology and understanding the technology. If individuals are not able to understand what it is, expose them to it so they can understand it, know how it works, know how it can impact their lives, and know most importantly how it works for business. The Why is the skills, the knowledge, and the understanding, which will lead to the implementation and growth of the technology.

Virtual reality is a job-disrupting technology. It allows individuals to work remotely. In one case, it has allowed an engineer in one area of the world to actually direct a technician in another area of the world. So obviously, that engineer could potentially manage several technicians in various locations around the world without having to actually be there, and the business would save money by hiring one engineer who can manage three or four technicians who are paid a lower salary. This can also be seen in the medical field where a highly skilled specialist could guide another trained physician virtually in a procedure instead of that person actually being there. We see that in logistics with drones being flown virtually, not directly, to deliver packages. We also see it with social media communications, with virtual chat rooms in which a person can communicate with various people from different areas. So it is important to understand that this technology is not only disrupting but displacing jobs in other areas. Instead of a person actually being physically there in a classroom, workshop, or demonstration, the individual can either virtually view a computer-generated scenario, view that person virtually, or practice and experience whatever the task may be virtually without having to go through role play or have an actual human interaction with a trainer or another colleague. This can eliminate costs for training periods. When we talk about education, we can think about online classes or virtual classes where now, instead of going through a learning management system, you can have students virtually in a classroom, joining with live students and interacting with the professor. It'll look as if they are in the classroom. They're able to talk and communicate and answer questions, and yet they could be anywhere in the world and do this virtually. They can be part of the classroom and never actually physically attend the class. Everything could be done in real time, and they could have the overall impression that this is being done in the present.

History of Virtual Reality

The leading edge of virtual reality can be traced back almost 100 years. We can see the relatives of present-day virtual reality when we look at items such as the View-Master or other attempts at putting virtual reality or enhancing 2D experiences. Virtual reality as we know it today was founded by Palmer Luckey with the creation of the Oculus Rift headset and subsequently the HTC Vive headset. These are the true ancestors of what we know as present-day virtual reality. All the main virtual reality products are either copies of or developed versions of the Oculus and the HTC Vive headsets. Also relevant are the Google Cardboard add-on and the 360-degree video, which is more widely used and considered virtual reality.

Today when we talk about virtual reality technology, we're actually talking about a threefold family of technology known as X Reality (XR) technology. XR technology is comprised of virtual reality, augmented reality, and mixed reality. The most common of those three is augmented reality that we commonly use with

Pokémon Go, Snapchat filters, Instagram filters, or Facebook filters. The reason they are commonly used is that they're easily accessible on smartphones and very easy to use.

There are two main types of augmented reality: (1) marker-based augmented reality and (2) non-marker-based augmented reality. The most common one is the non-marker-based virtual reality, which are the filters we use for marker-based augmented reality when you have to have some type of target that will trigger the experience. Often there are applications such as MAKAR that allow you as a business owner, for example, to have your customers go to the application, scan food or an item on a menu, and then a video pops up on their phone. There's something you have to do to trigger that experience, whereas with a non-market base, you can create the experience of immediately no trigger to activate it.

When we think about virtual reality, many of us think about games and headsets. The biggest market for virtual reality applications is the pornography industry, and second is the gaming industry. So it is correct that the gaming industry is one of the biggest markets for virtual reality, and it's growing very rapidly. The use case for virtual reality seems to follow what is called "Occam's razor"—when you have several choices, go with the simplest one. With virtual reality, the easiest and most simple use cases are 360-degree videos. These videos are easily accessible on a desktop computer, a tablet, and, more importantly, a smartphone. Many people don't consider 360-degree videos actual virtual reality; however, it is the most widely used virtual reality due to accessibility. Since the end user does not have to have a headset, you can actually engage with the content in the video very easily from your smartphone or any Internet-enabled device and build on that experience by having anything such as a Google Cardboard headset that allows you to scale your experience and also allows for scaling up accessibility of content. Whereas if you have content that is created for the Oculus Rift or the HTC Vive, you can't scale that content to your smartphone to view it and engage with it; however, you can scale a 360-degree video and view it if you go online on the Oculus Rift or Oculus Go website and experience that content in a VR headset.

One of the reasons I wrote this book is to talk about scalability of accessibility. I think many times as VR enthusiasts, hardware producers, and software producers, we want the end user to meet us; however, the reality is for practicality and scalability. We need to meet the end user, and at the time of the writing of this book, the majority of end users are using their smartphones to access the Internet. If we are to follow the success of augmented reality, it will have to be via the smartphone and the ability to scale up as the price point and the constraints that are around the current hardware diminish or as the hardware devices merge.

So the technology begins to move and the VR hardware begins to move closer to smartphone technology, and smartphone technology moves more to VR and becomes more integrated. Then we will be able to see more of a movement to the separate hardware. But otherwise, we have to go where the end users are. When we talk about education, we have to find ways to meet our students and expose them to the technology that will only develop and allow them to be able to scale the knowledge they already have, the grounded experience with the technology.

We don't know what it will look like in four or five years. As we mentioned before, right now Oculus Quest did not exist two years ago, let alone four years ago. It did not exist. We did not have that option. As for preparing our students, that exposure will help them move forward to understand the technology and at least give them some type of fundamentals and foundation.

As we move forward, virtual reality technology is

becoming more efficient. It is moving toward portable, easily accessible devices that are not very intrusive. Right now, the release of Oculus Quest is actually bringing a higher level of portable virtual reality and accessibility without having to be tethered to a gaming computer. That has brought down the cost and the disruption of the technology. Hopefully, that will allow for more users of the technology. The hope is that with a lower price point, there will be more adoption of virtual reality technology. It seems to be the trend that as the price point goes down, the quality and the restrictions go down. It becomes more open and more accessible, and you have more individuals buying the units.

There also has to be some type of connection or use case, and I think it ties back to what we discussed before as far as most people using their smartphones. So now you have your smartphone and your VR device, and there needs to be some type of connection or movement from one device to another. It could be something like instead of having a TV in your home, you have an Oculus or HTC headset and use that to watch TV. There has to be some type of connection because what we're moving toward or saying is that we're going to utilize one-third of the devices. In other words, instead of having our smartphone and HTC Vive, we're going to utilize our smartphone, TV, and VR devices. I think that as things are moving to more and more people, you can watch TV or live stream on your smartphone, so things are merging and becoming simpler. In the United States, STS in hardware designers are asking people to complicate their lives by adding a different device, so I think we need to move toward things that are streamlining the whole adoption of the technology, and I think that is what the future will hold for virtual reality.

Since the invention of the Oculus Vive by Palmer Luckey, there has been rapid development and advancement of virtual reality hardware and technology. We've gone from headsets that need to be connected to a gaming computer and motion sensors to the development of Oculus Go, and now we have Oculus Quest that bridges the gap from the gaming computer system and the motion sensor to having one stand-alone device. But will this rapid advancement continue, and will it be relevant to the consumer market and the mass market and move out of the early adopter stage?

There have also been developments with Google in the Cardboard system, which allowed easy access to videos on your smartphone. If we think about the history of virtual reality, development in the last seven years or the last decade has seen the greatest and most rapid growth and development in virtual reality technology as well as adoption. It is unspoken that pornography is the largest market for virtual reality, followed by the massive and rapidly growing gaming industry.

Virtual Reality Basics

Virtual reality is a computer-generated environment, so it can either be created with computer graphics (CG), a computerized environment, or real-life pictures and videos. There are different levels of immersion with virtual reality. On the lower end of the spectrum known as 360-degree videos, real-life photos or videos can be created that allow you to move around in the scene. However, you're not able to interact with anything in the environment. When you get to the computer-generated content, you're able to move around and interact in the environment. Also, the level of immersion is deeper, which means you have more of a realistic experience than when you're actually in an environment and able to move around, engaged and touching things. So you have what they call physics. You're able to move things around, pick things up, and actually immerse yourself in this environment.

Virtual reality can be accessed through a headset, a smartphone connected to a headset, or directly from your smartphone. The headset gives you the higher level of immersion, and then it goes even deeper, depending on what type of headset you have. The more sophisticated headsets give you even more levels of immersion or feeling that you're actually part of the environment in the engagement, and the realism of the scenario and the environment is greater.

Virtual reality uses apps similar to the smartphone in order to access experiences. For example, if you have an HTC Vive headset or an Oculus headset, you need to download apps. Many of those apps are free, and others you have to pay for. Those apps are what allow you to play games, interact with others, and have access to educational content. For example, there are apps you can practice interviews with. There is an app for Netflix, so if you have a Netflix account, you can access Netflix VR Hulu. You can also use your Oculus access or your Facebook account to communicate with your Facebook friends who are also online in Facebook.

There are also sports channels on which you can watch sporting events or concerts—a one-stop entertainment hub from which you can do the same things you can with a TV and your virtual reality headset. You can communicate with others, so it's a social device as well. When you think about it, instead of having a TV, you can actually have your virtual reality headset to carry around and use to communicate with others and watch movies.

Most people nowadays use streaming services, and you can access your streaming services with your virtual reality headset, talk to others, message people, or even surf the Web. It's a self-contained TV and a communication device that you can actually carry with you and access wherever you go. The only issue is with the price point. I get the cultural aspect of it. I like to use the analogy of Uber, which started about 10 years ago from the time of the writing of this book. It was not widely adopted or culturally accepted to get into a car with a stranger and have them drive you around. It wasn't culturally acceptable to ride in a car with other strangers and ride to your destination. Now it's normal to take an Uber anywhere in the world, and it's pretty much universally and culturally accepted.

With the virtual reality headset, currently it's not culturally normal to sit in a public place wearing a headset or using your headset. Often, the headset is viewed as a game console or game item that you use at home, so it's not viewed as something you want to take out in public and use in front of other people. I think that aspect of it will need to change.

Virtual reality is a game-changer, and the most widely accessed virtual reality is done on a smartphone through the use of what's known as 360-degree videos. As we move forward, there should be more adoption of virtual reality headset use. With the advent of 5G technology, virtual reality will have even better quality higher resolution in the near future, which should increase its use.

State of Education

My background is in higher education. As we talked about earlier, there was a catharsis, a moment at an academic conference of accounting professors in a presentation by our President Mark Rubin explaining how the future of the accounting profession really is in our hands. It is dependent on how we work to train our students, and right now, I see that many institutions are not fostering the innovation and technology needed to train our students. One thing we have to realize is that what we're training our students for today will potentially be obsolete in the next two to four years, and it's imperative as teachers, as educators, to stay ahead of the curve and push the envelope. It's important that we hold that door open for our students to go through, even if we only give them a glimpse or some skills they will need to have some type of foundation when they graduate.

So when I think about the state of education, I think about virtual reality technology. There are so many students in higher education who don't even understand what it is, how the applications work, and the different types of technology. Having that foundation will at least set them apart or allow them to move forward in four years because the technology is not going anywhere. It's only growing, developing, and evolving, and if you don't even have a basic understanding of what it is, let alone know how to use it and understanding the use cases, you won't have a leg up over a vast majority of students today. As educators, we have a duty to make sure our students understand the fundamentals.

Think about the superstar basketball coach John Wooden who focused on the bare bones fundamentals of the game. I think that's important because at the end of the day when we talk about virtual reality, we talk about virtual reality in the future and how it will look. We don't know, however, that the basic principles of it will be the same, and exposing our students to how it's impacting business today will give them the foundation and allow them to project, move forward, and see how it can scale, how it will be utilized in the future, and the impact it will have.

This is our duty, and the state of education is not equipped for virtual reality technology or technology as a whole, which is creating a rift. We have institutions, universities, and professors who are very progressive in their movement and adoption of technology. Other institutions are falling behind. I'm not saying that we need to spend millions of dollars on technology labs or VR labs. It could be as simple as awareness of the technology, and for educators, that means we have to learn and understand what it is in order to discuss it with our students.

There are definitely going to be students who are way ahead of us and have the skills, knowledge, and understanding of this technology. That's what People want. The reality is that when you sit down for an interview with an employer, you need to be ready to go. There's no training, and they want the smallest learning curve as possible. So we at least owe it to our students to give them the advantage or at least the best we can possibly give them so when they graduate they are able to compete with the students who have been working with technology from square one

Educators need to constantly be learning new technology. They don't need detailed knowledge of every new virtual reality update or application or software. But they do need to stay current and have an idea of what's happening with virtual reality and XR technology right now. They need to have an idea of where it's going because if they don't understand the present, they will have no idea or inkling of what will come in the future.

If we're always trying to catch up, it's not going to be a detriment to us; it's going to be a detriment to our students. We need to at least get our students up to speed with the technology we're using today so they have a strong foundation for the future.

The best option for educators is to stay abreast of the present so we can understand what the future will look like or at least move students in that direction or to an area that may have growth potential. We don't know what the future will be, but we know that it's rapidly changing, and because it's rapidly changing, we will know quickly if we're moving in a viable direction. We'll know if we move in another direction if that's a viable direction or if that technology would take off. If we move to an area that may not take off, it's not lost because something is learned from that technology that may be used in other areas. So it's important that we stay current and focused. It could be from professional learning opportunities, going to blogs or podcasts, taking free courses to understand and learn a current virtual reality technology, or paying for courses to go back and learn the technology. We need to be immersed in it so we can at least do the best for ourselves and our students.

There is a huge divergence in education in the United States when we talk about students and homework. Most students are assigned homework that requires them to have broadband Internet. Why is that important? Right now, we're implementing 5G Internet in certain areas of the country, and soon the broadband access that a large majority of children have will be outdated. Why is that important to virtual reality? In order to use the headsets at home, you need some type of Internet.

Many of our educators today are digital immigrants, born before the age of the Internet and the age of the smartphone. A majority of our students are digital natives, born with the Internet, smartphones, and technology. Teachers who are digital immigrants may be forcing their learning styles on the digital native students who have no idea what it means to read a book, go to the library, or search through a card catalog. These are new concepts for digital natives and are not relevant.

We have to remember that in the classroom we need to focus on innovation and change. Everything right now is in a constant state of change, so we need to focus on collaboration and students updating their skills, not necessarily memorizing or learning everything students need to know in order to find information, disseminate it, and develop skills. Right now, everyone has access to all the information they need. A recent article I read stated that on our smartphones we have access to more information than the President of the United States did 20 years ago. So we have to put that in relevant terms and understand that we live in an age of digital natives, and those digital natives have more certainty. In other words, they don't need to go to the library to look something up. They google it. They can look on their smartphones and gather information. We need to be cognizant of this and understand how education has evolved to today and how it will continue to evolve.

We also need to understand how virtual reality fits into this, and we need to understand it's not an up-and-coming technology. It is a technology that is here today, and students need to have that exposure, not necessarily that they have to know everything about virtual reality today, but they do need to be exposed to the technology so they can understand it and have the basic skills and foundation with the technology. That's the important thing right now because things are continuously updating and becoming new, so it's not the situation that digital immigrants are used to. If we think oh, I buy this smartphone and I'll have it for the next five years, or I buy this program and I'll be able to use it for five years. If I buy a smartphone now, it's probably good for less than one year and then a new model comes out. I buy software, and it may be obsolete in six months. So

I may not be able to learn all the functions, but I know the basic parts of it, and when it updates or when there's new software that may displace this one, some of those same functions and uses will still be there. I'll have the skills to navigate that new software and be able to easily learn the new features, the new add-ons, or the nuances to the other software that are there.

We are in a constant state of change, flux, and innovation. Everyone, not just educators, must continually self-invest through professional and personal development. We must take every opportunity to learn something new—new skills and concepts. The skills that we need are constantly changing. Virtual reality as we know it now will not be the same in the next few years. We must be flexible. We must adjust.

We also need to develop ourselves. We can do this through professional development sponsored by our employers. Many times we take for granted the power of personal development, which can come in the form of watching YouTube videos, taking free online courses, listening to podcasts, and going to meetings. This is very powerful because it seems that knowledge travels slowly. In other words, by the time we are able to take a college course in virtual reality, the technology may have moved years and years ahead. The biggest value in my opinion is from personal development, particularly events such as meetings. I've had the opportunity to meet individuals who own virtual reality companies. They are in the business and can give me specific details about how things work firsthand. They are at the cutting edge because this is their business; this is their livelihood. They have to study it, know it, and understand it.

Leveraging personal development as well as professional development is important. Personal development can be tailored and is self-directed so it's guided toward your interest level and skill level. Professional development is more standardized and is also highly valuable. For example, I may give a talk on virtual reality. Yes, I'm an expert in the education field and know more than maybe many about virtual reality, but I'm not working in virtual reality. That isn't my space. Virtual reality is a broad space, and that's something to keep in mind.

It's important to learn directly from individuals who are in that space so you understand their challenges and the nuances from A to Z. That can be done either in person by going to meetings, joining associations like the VR/AR Association, going to conferences and conventions on virtual reality, and listening to podcasts. If you can't talk directly to a person, listen to the podcast and go to websites and blogs. You may be able to ask individuals questions directly, and that contact will build up your personal development and help you gain the specific knowledge needed to really utilize virtual reality and understand the limitations, for example, in the classroom. You'll know that maybe you're not going to be able to get HTC Vive headsets for all of your 30 students, but there's still Google Cardboard or another solution, maybe even a merge cube. There's another solution you can use or app that's free that might not be 100%, but at least it could take you 20% of the way, and that 20% could be what's needed to at least bridge the gap from nothing. You can work with that 20% and work on filling in the other 80%.

Many times we believe that everything has to be 100% and everything has to be just right. We know that not everything is black and white, not everything is 100%. Our goal should be to understand the environment, understand the technology, and understand our situation and how we can work with our resources as educators and work with our students to learn about this technology. How can we bridge those gaps? Will the experience be the same now? Will there be an experience? That's what we need to focus on, the experience.

The other thing is to remember that we can't get the Oculus Rift or the latest HTC Vive, but in six months, a year, or two years, that technology may be obsolete. It may be dated when the newest version comes out. So even though you have the newest technology today, a year from now, that will be dated. There's always going to be these gaps. Some may be wider than others, but we don't want to have huge gaps to fill because that's challenging. The way to look at it is that you have one item today, and next year there will be a new item. There may be a gap of 10%, but how do you fill the gap of 80% or 90%. It needs to be filled so at least there's some type of foundation, some students who were working with the technology and can move forward, understand it better, and help improve it.

The state of education in the United States is abysmal. It is abysmal in the sense that we have not fostered innovation for our educators, and money is not evenly dispersed or distributed among the schools from grade school to high school. There are differences in the access to technology. There's uniformity with the curriculum, but it's a matter of access and exposure to technology, which is very skewed. There are some schools where students have access to the latest and greatest technology, and there are other schools where students do not have access to that technology. A majority of this technology is virtual reality access, which leaves many students behind.

Access to Virtual Reality

Access to virtual reality technology is not balanced or equal. We have the haves and the have-nots, and we have a case of early adoption. There are many who still view virtual reality only as a gaming system or something that little children use and don't see it as a job-displacing, disrupting technology in use today. There is access to virtual reality, but unfortunately, many who use virtual reality are considered early adopters. The reality is that virtual reality is hidden in plain sight.

When I have the opportunity to speak at conferences, many of the people have no idea what virtual reality is. They may say it is something for little kids or for gaming. They may say their son has an Oculus, but when someone asks them if they've looked at a new car or an apartment lately online or bought a house and see these things through 360-degree videos or through virtual tours, we begin to see that maybe virtual reality is actually hidden in plain sight. Access to full immersion virtual reality is a very distant reality for many people, especially in their homes. Many people may go to Dave and Buster's and other places for entertainment and try virtual reality, but there are few who actually use it in their homes and interact socially with others in virtual reality or use any of the apps that are available in virtual reality. Access is not available in the classroom and in education. It's even more of an abysmal situation when many educators have no idea what the technology is, and many times the students have more of an inkling or understanding of virtual reality than the educators. The disconcerting thing is that educators are the ones who are supposed to be teaching and showing the students things that are new and ultimately the ones who should at least pave the road for students to make a connection with the technology.

Many have said that 5G Internet will be a game-changer for virtual reality, and I believe it will. However, the majority of people will have Wi-Fi access and not 5G access when using their virtual reality headsets. When we look at the latest Oculus headset, the Quest, it is not 5G capable. I believe the practicality of 5G technology will not be realized for some time to come. There may be some companies that will begin to make hardware that is 5G-capable, but there are many households in the United States that do not have high-speed Internet connections, so it's far-flung that many of these households will have 5G when many of them don't have Wi-Fi connections in their houses. That will need to catch up first before moving to 5G, so there's going to be a disparity in those who have Wi-Fi technology or Wi-Fi Internet at home versus those who have 5G technology. And 5G technology is only in certain areas, so as that grows, the potential for hardware use will rise. But again, there will be a disparity. There may be hardware that is 5G, but if the majority of people have Wi-Fi, they're not going to actually be able to use those features that are 5G.

Access to virtual reality may always be an issue, and this is not necessarily a bad thing. The difference in access would be in the levels of hardware. Currently, the broadest or most accessed virtual realities through the smartphone or Google are the most common, and HTC Vive or Oculus Rift headsets are the least common. However, their numbers are growing, so I think the path of least resistance is where we find the most access and where we will see the most growth. There needs to be an emphasis on this area and how to develop and improve on it, which would increase the access to virtual reality.

People watch movies and TV shows, and they're usually not wearing headsets. They're wearing some type of device that connects to their brain somehow and moves them to a different place. This shows us the futuristic idea that virtual reality is less intrusive. My hypothesis is that there will be some type of integration, either a device that is like a smartphone that gives us virtual reality access or some type of link to the smartphone. If there's some way to give us easy, deep immersive access to virtual reality through our smartphones and it's simple to use, it would probably be easily adopted and it's not intrusive. When I say simple to use, it's not intrusive, which means you could walk around with whatever device and see things but also move from a virtual world to the real world. I think that would be a hit.

When we talk about access to virtual reality, it's not really about access; it's about economics. When I think about the classroom and how I can implement the technology or let students use a technology, it's not a matter of the technology not being available. It's not a matter of not enough content being available. It's not a matter of the hardware. It's a matter of having the money and the resources so everyone can use it. When we talk about access, we have to think about economics. We have to think about money and how it impacts access to the technology. This not only applies to virtual reality but also to other technologies because economics is a factor. In other words, as there are more and more iterations and the technology is either older versions or becomes less valuable or cheaper, or the technology itself becomes very cheap to make and ubiquitous, the point we have to look at is the dollars, the resources.

It's a community issue, and we have to think about how we want to use our resources, how valuable virtual reality is in education, how valuable it is in the workplace, and how valuable it is as far as an investment in the future. Many times we look at the return we're getting right now and don't think about the return in the long term. When we take the economic approach, we think let's wait before we buy; let's wait because technology changes so fast. The issue with that is you're not in the water, and you have to get in the water in order to swim. If you're not in the water, you don't have the opportunity to swim or move forward, so it's important, even if technology does change in the next six or seven months, to get in the water and swim.

When it comes to education, students are able to have an experience, an impact in some sort of foundation with the technology that allows them to move forward. So instead of swimming the breaststroke, it's easier to learn the backstroke because they at least got in the water. It's important to think about the economics, and really, that is the challenge with virtual reality. It's not the technology; it's a matter of resources, economics, and development, because when we talk about resources, we're talking about development and how we focus our development, how we focus our move forward. Many times, it's hard to change.

I just read about the gentleman who invented the telegraph. He went to the military to tell them about his amazing invention, how a person can communicate over long distances. The military said they didn't need it because the system of using flags worked perfectly. Eventually, of course, the telegraph revolutionized communication. The point is that many times we cling to the past or refuse to see the future. People are usually comfortable with change because change often makes our lives easier. Technology has brought us vast improvements, but the problem is not making connections. When I talk to people about virtual reality, it's hard for them to get the connection about how this will impact them.

When I'm using analogies, I'm reminded of an interview with Jeff Bezos, the CEO of Amazon. He talked about his friends and family fund raising. He explained that he wanted to sell books over the Internet. Everyone thought it was a great idea, but the question he got was this: What is the Internet? The challenge is to get people to connect the dots and understand how something works, how it impacts them, and how it will impact them day to day.

Access to virtual reality technology in schools is important because that exposure will lead to greater technology skills for those students who have access and exposure. We have to think in terms of exposure to technology and not necessarily mastery of technology because technology is rapidly changing, and mastering the technology may not be feasible. By the time students master a technology or we put formal education requirements on it, the technology will be obsolete by the time the students graduate or move forward in their education. We have to think in terms of exposing students to technology as much and as soon as possible. For example, with virtual reality technology, students need to be exposed to it. They need to gather the skills to use it and understand its impact. Then they need to know how it can be used in the future. There will not be that steep learning curve when they enter the workforce.

Right now, the workforce is based on skills, and having those technology skills is important. Virtual reality in data analytics programming skills is very important. Will programming languages be the same 10 years from now, five years from now? They could be. We don't know. Things could totally change, but there will still be programming. Will virtual reality be around? Yes. Will it be around in the same shape and form as it is now? No. We think about this 5G technology that's being adopted in the United States, and hardware manufacturers for virtual reality are not even making headsets designed to run on 5G. So in a matter of five years, we won't know what the technology will look like, but the exposure now to that technology is important. If you have a student who's a freshman, by the time that student graduates, 5G will be widely adopted and it will be a totally different world. But that student has the exposure and at least a basic knowledge of the technology and can move forward with it in some shape or form.

What Do the Zealots Think?

When I talk about zealots, it is not a negative term. It is in reference to the early adopters, the innovators, those first movers, the embracers of virtual reality technology. These are the individuals who developed the hardware and the software, who developed apps to start companies and businesses around virtual reality from the onset. It was also those who made the first purchase, the latest headset, the Oculus, the HTC Vive, or the Oculus Rift. So this is not to be taken lightly; in a bad light, this is a good thing.

The challenge is to bridge the gap from these zealots, those early adopters and innovators. I consider myself part of that group of early adopters and innovators who see that this technology is the future. They see that this technology is very important and that it's highly relevant. The only problem is that the other 95% may not have the same vision, and that's not a bad thing. That's the reality of it, and we have to realize that we cannot bring that 95% to us; we need to go to the 95%. The reason I call this group zealots is because the mindset tends to be that our 5% group means to bring the technology to the 95%. That is why we receive so much pushback. That is why there is so much misunderstanding around the technology and what it represents. There has to be a middle ground, there has to be a compromise, there has to be a meeting point of what that will look like or how that will work. I can't say it depends on the space that you're in. What I can say is that from the technology or software company side, you have to flip to the other side, that 95%. For that 95%, there is no urgency around this technology.

There were actually two interviews with Jeff Bezos. In one interview, they asked him about his friends and family and the fund raising. He said his family and friends were very supportive and only had one question: What is the Internet? Then he explained in another interview how he had to raise a million dollars and meet with about 50 investors. About 22 of those investors gave something like $50,000 each. And all those investors asked him the same question: What is the Internet? For many educators, highly intelligent and knowledgeable people, the question is this: What is virtual reality? We need to do a better job answering the question about virtual reality.

When Bezos explained what Amazon would do, I believe he talked about the benefits—how it would help not only Americans but the world. I think that's the mindset we have to have—not so much a focus on the technology and the software but on the benefits and how it will help students, consumers, businesses, and individuals around the globe. We need to focus on the benefits, and in the end, we'll be able to answer this question: What is virtual reality? We will have done a good job if we can answer that question because those individuals we've talked to will see the value and the benefits. We want to leave them wondering about the technology and the benefits of that technology. Let's focus on the benefits.

The early adopters are always eager to have everyone use the latest and greatest technology. The reality is that only a small portion of the population will even be aware of this technology, let alone how to use the technology. This is the case when we talk about the newest technology or the newest programs that are available. They are often amazing and have a lot of potential. The problem is that many of the early adopters and innovators did not realize that the majority of people have not even heard of the most basic technology and functions, let alone the newest and most advanced versions of VR technology. So we have to take the visions and the thoughts of early adopters and innovators seriously, and at the same time, we have to realize what the actual reality is with the use cases of the technology. When we talk about 5G Internet, it is amazing because it's actually available now; however, only a small part of the population has access to 5G, which translates to very high-end virtual reality experiences. Also remember that virtual reality hardware is not a majority of virtual reality hardware and is not designed to run on 5G. It is designed to run on broadband speeds. We need to remember that there is a large number of households in the United States that do not have access to broadband. So when we talk about 5G speed changing virtual reality, we have to realize that the mass of people out there do not have access to most of this technology and will not see the benefits of it for some time.

It is important to have early adopters and innovators of technology. These are the individuals who pioneer and push the technology forward. They are the ones who troubleshoot, who forecast where technology is going. I consider myself an innovator or early adopter of virtual reality technology, and I understand how the technology works, where it will go, and how it will develop and move forward. So the zealots, as I call them, are very important and can see the whole spectrum. What tends to happen is that early adopters and innovators tend to be around other early adopters and innovators and may have tunnel vision. They may only focus on those who are using the technology and see the benefits of the technology. We need to think about the wide spectrum.

You have to remember that the innovators and early adopters are probably 3% of the population, and we need to focus on the other 97% of the population who may have no idea of the technology or may not understand how to use the technology. It would help to push virtual reality forward and move the understanding of the technology forward because we have to meet on middle ground. We might be in the black on the opposite end of the 97% who are in the white space, and we need to find that grey space in between where we can at least meet people halfway and get them to understand where this is going and how urgent and important it is.

Often when I'm at conferences, I know the technology being used and the multi-million-dollar companies built on virtual reality technology. Many times, I can get others to step into that gray space and begin to see and understand how this works, and that this is important. The only problem is that they don't believe it can actually be happening now. They believe that maybe in the next five or 10 years, they will see that application growing, working, and being used. The problem is that it is already being used by people making millions of dollars from it. But remember, the win is getting those individuals to understand the value of the technology. We see the value of the technology because we're the innovators and early adopters, and we understand how it's working. What needs to be done is to get people up to speed to understand that this is more than just a video game at Dave and Buster's. This is technology and skills that are being used to help businesses and society.

People need to understand the value that virtual reality technology brings. I think that's a win. It's moving from the black space to the gray space. It may not be pulling in someone completely from the white space to the black space, but it's a win because at least the knowledge is there, understanding has increased, and the value is there.

As innovators, we have to let that 97% understand the technology and see the value in the technology. If there's no value, then there's no way to move forward. Cost is only an issue in the absence of value, so if that 97% doesn't have any value in this technology, it doesn't matter if it's $10 or $20 or $2,000 for a headset; no one's going to want to pay that kind of money for it, and no one's going to want to purchase it because there's value to it. We need to focus on the value and not so much on the benefits. Let's focus on the value the technology brings, and then people will begin to see how important virtual reality is.

The Practical and the Tactical

We need to be able to meet educators and individuals at their level. We need to be cognizant of budget constraints when it comes to educators using virtual reality technology in the classroom.

It comes down to actually being honest and understanding what the landscape currently is in the environment. We need to take our time and actually find ways to slowly introduce the technology when needed, take the time to train individuals, and also take the time to understand what their needs are. Sometimes it may be using 360-degree videos on a smartphone instead of having a full-blown HTC Vive headset and a high-powered computer. It may mean not a totally full immersive experience that we want individuals to have. It would be tailored to their needs and tailored to give exposure, particularly in the educational environment where budgets may be tight and there may not be enough funds available to purchase headsets for every child. It may be the option of using Google Cardboard and free software that's available. It will allow students to experience virtual reality technology first-hand. It would be the same experience as an HTC Vive or Oculus Rift gnome, but it would be an experience. We should look at weighing the options between experience and no experience and also see and understand who we're working with and who we're dealing with instead of trying to impose our ideals and technologies on individuals. We need to bring them to us and understand where they're coming from and what's important to them.

The practical applications of virtual reality in the classroom are at zero. The current cost of virtual reality headsets on the high end of the spectrum is cost-prohibitive for most educational institutions, even though the technology has a strong impact on the work world in the educational environment. As educators, we have to figure out ways to work within our budgets and at the same time be able to give our students experiences with new technologies. For example, if we don't have the budget for certain applications or technologies, we have to let students share, or we may have to download free apps for students to use instead of using paid apps. If there's no budget for technology at all, the other alternative is to have thought experiments and have students think how they would use the technology. At least in their minds, they're thinking about how the technology can be used in various situations.

On the practical side, it is important to consider that there is a divide between those who have access to the technology and those who do not. You have those who have money or households that have more money and more income to pay for VR and broadband Internet in the home, which is the fuel for virtual reality technology. The other is that as we move into 5G, it is only in certain urban locations that tend to be affluent. Currently, the majority of VR headsets with technology are not made for 5G technology. They are still made for the masses that are using them on broadband Internet. But the point of this is to show that there is a divide.

So we have the haves and the have-nots, or those who are able to have the technology and those who are not. We have to keep in mind that there are a good number of people who do not have the technology, let alone exposure to the technology. We should be cognizant of that and realize that there has to be a happy medium. If someone could come in right now and meet the gap or meet the gap with the device, not necessarily with their price point but with some type of device that would bridge the smartphone and the headset, it would be a great opportunity.

We can call the smartphone-to-headset gap the cultural gap, and we can call the price-point gap the class gap. If there's a business that can come in and bridge those gaps, it would be very impactful. It seems that the strategy with Oculus is that they continue to develop and scale down their price point, so they move up in the technology and then lower their price point, but it's still not an optimal price point for the masses.

The Present

We currently live in a world where digital equity has a strong rift. There is a great digital divide when we think about our very tight educational system technology budgets as a priority. However, in many school districts, technology may not be able to be updated or training may not be available for educators to teach the students the new technology. This has to be a constant line item on budgets that may be limited, which creates a disparity for children.

When we talk about resources and household spending, technology, especially virtual reality, is not an urgent matter in many households. We talk about basic essentials such as laptops and Internet service, but when we think about virtual reality, we need high-speed Internet in the home in order to utilize the system effectively. When we talk about lower-end systems, we mean Google Cardboard or Samsung Gear that they can use on their smartphones using their data package. Presently, there is a digital divide or disparity between those households that have the support services that will facilitate and even use virtual reality. We think about the whole culture and the acceptance of using virtual reality outside the home and the convenience of it, so there is an adoption curve we have to think about. There's also a cultural curve and a financial aspect when we think about the services that are needed such as phone data plans for high-speed Internet to support the virtual realities. As we move to 5G Internet, there are many areas that do not even have high-speed Internet, let alone 5G, so when we think about rural areas and poor urban communities, these are realities of a digital divide and barriers to entry and acceptance of using virtual reality technology. The future does look bright for the technology, but it will be a steep adoption curve.

There is an economic divide between those who can easily afford virtual reality and those who cannot. The price point for hardware has been on the decline, which makes adoption easier. However, the price point has not come down so much that it can make the technology ubiquitous. We also have to remember that smartphone technology has been on the increase in price, and smartphones now cost more and we have virtual reality technology going down. Again, I believe there needs to be some type of convergence between the two technologies, some type of crossroads or hybrid or meeting point where the devices are one, so we have VR with our smartphones and we're able to purchase the smartphone and then use VR instead of purchasing two devices. Now we have to think that we have a TV, maybe a laptop or desktop computer, and then we have a smartphone. What's happening now is that we can use our smartphones as our TVs and as our computers, but now we have to purchase a VR headset, which is an extra item, and VR headsets can be our TV and we can access social media. So now we have two devices. The only difference is that the VR device is not as practical or culturally accepted as the smartphone, so we can carry our smartphones around with us at work, at school, outside the house, or in our cars, whereas the VR headset is not widely culturally acceptable to put it on at work, at school, or sitting in the airport. So these things cause a rift that needs to be addressed.

The present state of things is difficult to change. Normal people are not very open to change, and with virtual reality, it is more of a cultural shift than a logical, technical shift. We need to focus on how we can culturally get virtual reality accepted. What I mean by this is that if we think about Uber, a technology company, it's also a cultural company. For example, about 10 years ago, right when Uber started, one of my cousins told me she would never take an Uber. She did not feel comfortable riding in a strange person's car at night. If we think about the cultural aspect of Uber, it was akin to hitchhiking at that time and in the location I was living—Chicago. It was not normal to hitchhike or get in a car with someone you don't know and give them money. Culturally, it was not accepted, so there had to be things that were done to build trust and create a cultural shift so it would be normal.

I spent time in Oslo, Norway, where it was common on a Friday or Saturday night to take what they call the pirate taxi, or individuals with their own cars who would drive you around for a lower price than a regular taxi. You don't know these individuals, so you get in the car with a stranger, and in Oslo, that was culturally acceptable. With virtual reality, we focus on the technology, but we also focus on the cultural shift either by changing the technology or changing the way people think about it. That's really what's going to push forward the adoption and movement in the use of virtual reality technology.

Another example is the driverless car. That technology may not be where we would currently like it to be, but the cultural shift has to take place. Are people comfortable getting into a car that drives itself or being on the road with a car that drives itself? How are we going to shift culturally so it's not about the technology but about the psychology behind it? How can we change people's mindsets? Adopting the technology is really the challenge, and that's what's going to happen. Either the technology has to meet society or the culture where it's at, or the culture has to be cultivated and changed to meet the technology.

There has to be some type of meeting in the middle where culture shifts technology and they meet in the middle. I can't say how that will shift in the technology or the culture. But that has to happen in order for virtual reality technology to move forward at a rapid pace. It's the same pace as Uber when you think about its growth over 10 years. People were nervous about Uber, and now the word is as common as Google. It's how we travel somewhere, how we move, and how we shift cultures and move this around worldwide.

The Future

The future of virtual reality technology is very bright as we move into the age of 5G Internet speeds that promise almost real-time Internet experiences and will uplift the virtual reality experience exponentially. The adoption of the technology will more than likely increase as the price point of the technology decreases. More people will own headsets and heavy actual technology. More schools and educational institutions will be able to budget forward to have virtual reality technology. More private businesses will be able to have virtual reality technology due to the price point adjustment and the scaling up of the technology.

There will be more of a cultural acceptance of the technology as it becomes more normal to actually wear a headset in public, whether you're in the classroom, on the bus or train, or waiting for your airplane. It will become more accessible and acceptable to utilize the technology. There will be a strong cultural shift as well as an emerging or pairing of VR technologies with existing technologies. There will be some type of connection with your smartphone and your VR device so you can move freely and have contact with SMS messages or emails. You'll be able to view all of that from your smartphone so you won't have to leave the virtual world to check your smartphone or notifications or answer your phone. I believe there will be a merging of the two technologies to make them both very efficient. There will also be an influx of augmented reality uses where you're actually able to see the real world or what's going on. That feature will also be added to VR headsets where you will be able to see what's going on and move back and forth from the virtual world to the real world. You'll be able to get real updates and communication, take phone calls, view emails, surf the Web, play games, and watch movies all in your virtual environment.

The driver for virtual reality technology and usage is actually the cousin to virtual reality, or augmented reality, which is in widespread use throughout social media. It is easily accessible since it is a smartphone and has continued to grow in its application use. It is also not intrusive and is very social. In other words, you could play games like Pokémon Go or use filters to share. With virtual reality, you have a headset and are taken away from your environment or other individuals. That is not to say that virtual reality is not as social of a technology.

It is social in its own environment, whereas augmented reality is more of an external type of technology where you can use it with others. It is easy to share, and it's part of your own environment, which I believe will add to the growth of augmented reality. That does not mean that virtual reality will not grow; however, the more augmented reality is used, the greater the chances for the cultural shift in the implementation of virtual reality.

Because virtual reality is one step above augmented reality, instead of actually interacting with others or objects that are in your environment, you can actually go to another environment and interact with them. So virtual reality takes you to another level. The more we use augmented reality, the greater the chance of adopting it and increasing the usage of virtual reality. Also, the more people are aware of what the technology is, the more they will use the technologies today, even though they have no idea what they're using or what the implications of it are and how it's connected to other forms of technology. For example, the average person probably has no idea how augmented reality is connected to virtual reality and that virtual reality is connected to mixed reality.

It is thought that 5G technology will revolutionize virtual reality. The only issue is that currently 5G technology is not widely adopted in the United States. The cable companies have monopolized the wiring in the United States, and the cellular companies are the ones who actually use the towers for signal and for Internet. We have yet to see how well the adoption of 5G Internet will go in the United States, so if there is not widespread adoption of 5G, then it makes no sense for virtual reality hardware makers to create hardware that runs on 5G when no one has 5G. What is interesting is that countries like India are on 5G technology, using it however the hardware is designed for places like India and not the United States. It doesn't seem like the revolution of virtual reality technology will happen in the United States. It may happen overseas where there is 5G technology.

Definitely 5G technology will make virtual reality technology run at a higher level and at a higher speed and higher resolution, just like moving from 3G to 4G. But how long will that take to happen at a sustainable rate in the United States? It would make sense for hardware manufacturers to move forward with producing technology for that period. I think that's going to take some time to happen, and then when transitioning to move, there may be some devices that run on both 4G and 5G, but the time it will take for devices to be created to only run on 5G will take a considerable amount of time. That's not to say that 5G doesn't exist. It does exist in the United States, and some have access to it. As I mentioned before, the number of people who have access to 5G technology does not warrant creating hardware for them when the masses of people have 4G technology. The other thing is to remember that not everyone has access to 4G or high-speed Internet, so when we look at rural areas and even certain urban areas, there's no access to 4G Internet. These are things that need to be considered as we move forward and look to the adoption of the technology.

The Digital Divide

The digital divide can be defined as the uneven distribution of technology among geographic regions. It could be urban or rural; it could be regional through countries; it could also be through class or income-based disparities of access to technology. In the United States, we hear more about rural versus urban as far as access to the Internet. When I say Internet, I mean high-speed broadband Internet. We also hear about the homework gap where you have students who do not have broadband access or high-speed Internet at home. They are not able to do their homework on their computers or tablets, so they either have to do their assignments on their smartphones or go to McDonald's or Starbucks or a location that has free Wi-Fi. That creates what is known as the homework gap.

When it comes to virtual reality, the digital divide is in access and awareness, so when we talk about access, it is being able to use the technology and being able to interact with the hardware and software. This is very important because when we talk about the three main disrupters, there's virtual reality, there's blockchain, and there's artificial intelligence. When we look at those three technologies, virtual reality is the only one that is in use today. It's disrupting right now. We can use it for gaming if it's for entertainment, or we can buy an Oculus Go or Oculus Quest headset to watch movies on Netflix, play games, or go on social media. All these things exist.

There are companies like Next/Now virtual reality advertising agencies and multi-million-dollar companies that provide virtual store modeling. But all of these aren't even the tip the iceberg for the medical field and the use of virtual reality for training in surgery. When we look at blockchain and artificial intelligence and read about them in journals, three years from 2019 there will be a disruption with blockchain and artificial intelligence for years to come. That is not happening right now with that technology, but it is where the urgency comes with virtual reality and exposing students to it. I emphasize exposure because when we have a student who's a freshman in college, they're going to graduate in four years and go into the workforce. In four years, the technology will be totally different than what it is right now, and we need to expose those students and give them at least the foundation to see what the technology can do and understand it so they have the basic skills or fundamentals as they move forward. They can build on that and understand how these use cases are in the work world. The best case scenario is for the students to be able to understand everything from the hardware to the software, but that may not be the case. There are always limited resources that create limited availability to give students access to hardware and software; however, there are ways to give exposure, so it may not be the optimal condition. But there should be at least an attempt to expose students to this technology so they can at least understand what it is. It shouldn't be a situation where you talk to a student, a young adult, or someone else and say, "Hey, have you used VR?" They don't understand what it is or how it works, and they can't explain the impact, and that is a divide that urgently needs to be bridged.

When looking to bridge the digital divide, I believe there are three areas. The first is honesty. Having honest conversations with yourself and assessing the situation if you're an educator means understanding that if you have a budget, if you have resources to incorporate virtual reality, you need to know what level of virtual reality you're able to incorporate. For example, if there is no budget or very little money or resources that can be allocated to virtual reality, then it's a matter of thinking of workarounds, thinking of using 360-degree videos and free applications that can allow students to have access to virtual reality. Or it might mean having thought lessons so students can think about how they can use the technology, even if there is nothing available for them to use.

The second area is to foster a culture of virtual reality, which is fostering the culture of innovation. We have to begin to think about incorporating the technology, learning about the technology, and understanding how it will advance. That is the most challenging part. It has to start with small conversations. It can start with baby steps and getting to talk to people so they can learn more about the technology and simply move forward. It's been said that football games are won by inches, not yards.

The third area is reevaluation. I've heard many times from friends who own virtual reality software or hardware companies that many organizations are hesitant to make a purchase because they're worried the technology may become obsolete and that they will waste a great deal of money purchasing equipment. Within a year or so, they believe it will not be useful because new models will come out. That goes back to the culture. We have to realize that technology is advancing faster than we can keep up. The strategy in conversations around the resources have to be made to decide what to purchase or what not to purchase and what is important and what is not important for our institutional organization. These are the conversations we need to have. Maybe it isn't wise to make a one-time purchase of hardware. Maybe it is more efficient to purchase some hardware but also allocate funds to make purchases every year in order to get the most up-to-date technology. Or maybe it doesn't matter that a new version of a headset will come out next year.

The point is to utilize an exploded technology that's there, so it really comes down to the strategy, the budget, and having these conversations. It's also about going back and seeing if the strategy you're using makes sense. It may make more sense to make a one-time purchase, or it may make sense to allocate the funds over a five-year or 10-year period in order to have the continuity of new technology or applications. This should all be part of the strategy of the organization when it comes to virtual reality or innovation in general. It is very important and it's done frequently because technology changes frequently.

Technology is reportedly compressing speed for computer updates every 18 months. It is faster than that, so it should be done on a quarterly basis or at least on an annual basis to really understand what the organization's needs are and how they've changed. Even if it's three months, things change with technology, and there are new advancements that could help or that we need to be exposed to, and there may be even greater advancements in the next year or so.

We need to first be honest with ourselves and understand our situation. If we're honest and understand our situation, we're able to know what we can do and what we can use. We might be able to purchase Oculus Go headsets for all of our students, or maybe we need Google Cardboard for students or nothing at all and can actually view 360-degree videos to give students access to virtual reality.

Second, we need to work on the culture of the organization or the school that we're working for and foster innovation. That can start with simple conversations with our colleagues, coworkers, friends, or bosses about ideas we have about virtual reality.

The third is to reevaluate, to look at what we've done, how we've done it, and how we can improve on it. That's the important thing. Since football games are won by inches and not yards, if we take the time to reevaluate and reassess, we may be able to make changes or make what we have that much better. It may take longer, and in some cases it may take months. In other cases, it may take years, but by reevaluating, we're looking for ways to progress.

As we look at the developments of 5G mobile technology in the United States, we see that there is a disparity among countries that have 5G technology, countries that are developing 5G technologies, and countries that don't even have access to 5G technology. When we look at the United States and the European Union, we're pushing forward with 5G. In China, which claims to have 5G networks, and in South Africa which has tested and is considered to be using 5G technology, there's a great disparity. The disparity is between African and developing nations that do not even have access to the technology and are still using 3G and 4G technologies.

It is time to move forward, and it will take these countries time to catch up to the rest of the world and the technology. That means that again there will be less exposure to that technology and the benefits of it. From the domestic side, we have to think about the United States and those areas that do have 5G technology. We have to think about the exposure and the ability for those that have it to use it and then those that do not have it. When we look at 4G technology or broadband Internet, there's still a significant amount of the population that does not have access to broadband in the United States. More than 30% of rural Americans do not have access to broadband Internet.

When we think about the advent of 5G technology, how will that impact those who don't even have broadband? And how many others will not have access to 5G technology, which is a higher speed and provides more access and opportunity to exposure? As we can see, there are disparities. How can we prevent them, and how will we be able to react to the impacts? In other words, we may not be able to stop the development of new technologies and stop the disparity, but what can we do to lessen the impact and know that it does exist? I sense there is a disparity, and what can we do about it?

As for virtual reality, we know that headset makers such as Oculus have not planned to make headsets that are 5G-compatible because most people do not have 5G. So at this time, there's not a large impact on virtual reality; however, when we think of our smartphones, Samsung launched the S10, which is 5G-compatible. How will that impact VR on the phone and augmented reality on the phone? Will having access to higher speeds create an opportunity for further development with VR using your mobile device? That is yet to be seen.

Technology Anxiety

Have you ever felt like you don't know what's going on? Have you ever felt like things are passing you by? Technology is advancing and developing faster than anyone can imagine, and it's only increasing. At one point, you buy a smartphone. You can use that phone for several years. You get used to it and then get another model. Right now, there are new smartphones or models of phones coming out in less than a year. Those models are faster, take better pictures, and can do more now that we have 5G. When we talk about the world of virtual reality, we have to think about a year ago when we had Oculus Go. Today we have Oculus Quest. What will we have next year? With processing speeds doubling every 18 months or less, and when you think about Internet speeds getting faster and faster, it is very hard for us to keep up, and it creates anxiety.

The underlayer of this anxiety is change, constant change. To really grapple with this anxiety, we need a twofold understanding that it has changed and that it will not stop. First, we need to accept that things are constantly changing and evolving, which means we have to look at ourselves and realize that we're changing and constantly evolving as well in order to keep up. Otherwise, we become obsolete and we're not able to function. Eventually, if we don't update the technology we have, we won't be able to interact with others in society. So we have to keep up. We have to have the mindset of constantly evolving and advancing. That doesn't mean we should be innovators or early adopters, but we should embrace that this is not going to last. It will be temporary, and when it comes to virtual reality, we need to understand that the technology itself, the hardware, is constantly changing, and so is the software. However, the concept and skills that are needed to use it and develop it will stay constant, and that's what we need to understand and utilize.

It is a misconception that digital natives are more adept in technology than digital immigrants. Digital natives have grown up with technology and have been using it; however, that does not mean they are more adept. Digital immigrants have been using technology as well and have to use it to survive. They have to change in order to move forward. The next question is this: How do the digital natives use technology? Are they using the technology to build professional skills? Are they using it to gain on social networks? And when they're not using technology, how does it make us anxious?

We're so used to being connected, or what they call passively attentive, which means checking on social media, reading, and doing other things. But we're not actually multitasking; we're actually bouncing from one thing to another just to stay connected, and if we're not connected, it creates a form of anxiety because we don't know what's going on in social media. We don't know what's happening in the news, and we don't know what's going on in the financial markets. We haven't seen our text messages, and that creates anxiety.

With virtual reality, the anxiety is the same, but we're removed from our other devices. We're removed from many of our messages, text messages, and other things on social media because we're in a virtual world. When we're out of the virtual world, we're not able to interact, play games, or be on social networks virtually. That can create anxiety to not be able to connect virtually with what we've been doing, so there's that separation that virtually creates anxiety and tension among us.

I believe that virtual reality can potentially create a form of technology anxiety because in the virtual worlds, we can enter into chat rooms, chat with people around the world, or hang out in spaces or play games that are maybe more exciting to people than their actual real lives. The more connected we are on our smartphones and virtual reality technology, the more disconnected we are from actual face-to-face, in-person interaction and communication. We're able to virtually hang out with friends instead of actually being with them in our homes or at a café. We have virtual rooms where we can meet with them and play music and enjoy, so imagine if you don't have much, if you don't have a video player or a TV or your friend is not nearby. It may be more fun or more of a social connection to be in the virtual world than the actual real world.

We also have to remember that we can travel to different places in the virtual world. We can also create avatars that may be more attractive in our opinions than our own selves, so we may actually have some type of anxiety about ourselves when we're outside of our virtual world. There may be people who enjoy being in the virtual world more than being in the real world. I know from my own experiences with the Oculus Go that I can be in there all day spending time watching Netflix. It looks like you're in a theater, which is a lot better than my living room. Or I can go to a sports game. I was watching a Netflix movie and then left to watch a hockey game. Then I went into one of my social rooms and played music, and from there I played a shooting game.

You can have all these experiences and all this entertainment and travel. When I get out of the virtual world and I'm back in my living room, it's not very exciting.

So you can have anxiety from that. That's something we need to consider. But how long should we spin in virtual reality, and how will that impact others? I was able to go in a chat room and talk to people around the world, and when I got out of that, I wasn't able to do that in my own home or in the real world.

Technology gives us the ability to eliminate uncertainty so we can figure out where to travel using Google Maps. We can look for potential mates through apps, which creates certainty but also gives us uncertainty. How will virtual reality impact our certainty and uncertainty? Will it give us an escape? Will it allow us to actually experience things without actually experiencing things? For example, I like the application where you can practice an interview. You actually sit at a table during the interview, or you're standing giving a speech or presentation. Will this replace the actual uncertainty of being in front of a live audience or being in front of an interviewer? It does help by giving you some practice, but will it take away the uncertainty? That is the question.

Does virtual reality take away uncertainty? With many tasks, virtual reality allows us to actually practice or see how things are before we actually experience them in real life. The application that allows you to practice interviews and practice giving speeches allows you almost unlimited practice. It tracks your word count and lets you know if you're talking too fast or too slow or whether your eye contact is right. It can take away the uncertainty of an interview in real life, or maybe it will create more uncertainty because we're not able to actually cope with a real life interview. What if something goes wrong? What if the dynamics change? What if the person stands up instead of sits down? What if the questions take you off guard? The virtual reality app may actually give us more uncertainty. Being able to practice unlimited times and being able to have eye contact to understand your positioning when you're giving a talk or an interview is great, but does that take away from our people skills?

A great example of this is Google Maps. Before Google Maps, we would either go on our computers or go on MapQuest and either put in a destination, print out the directions to get to where we needed to go, or even, God forbid, buy an actual map. Now with certainty, we have Google Maps on our smartphones. It gives us directions and tells us how long it will take to get there on the fastest route. We don't even need to look to see where we're going because we can set the preferences for tolls or no tolls, and Google Maps will guide us along the way to turn left or turn right and tell us which roads to take. It even tells us if there are speed traps ahead. So this takes away a lot of the uncertainty of getting to and from point A to point B. We don't think about it. It doesn't matter where we are, what city, or what country. We can figure out how to get around, how long it will take with a rideshare, walking, public transportation, or in an automobile.

The uncertainty in this is that if we don't have Google Maps, what do we do? How do we navigate? How do we figure out how to get somewhere? How are we going to get there because we don't have Google Maps? If we don't have our smartphones and don't have Google Maps, we're going to have a lot of uncertainty about how to figure out how to get around because we don't think about it. We don't think about having to figure out landmarks and remember where we live or the address of where we are going. Everything is automated. We just do it.

So the question is this: Will uncertainty with virtual reality be taken away only to bring us back to reality and use those skills that lead to people in the real world? Will that become more valuable just like it is valuable right now to know how to get from point A to point B without using Google Maps? That would be highly valuable if you don't have your smartphone. Will the skill of dealing with or working with individuals in the real world become more highly valued as virtual reality becomes more widely adopted? And here is the next question. Will virtual reality with all the algorithms and programs be able to mimic the spontaneity of the real world to a point where the virtual and the real are one and the same thing?

If we think about it, there's a lag, so when I go into that interview program or that program that allows me to practice speaking on stage, there's not that many variables of spontaneity. What if I trip? What if there's bad lighting? What if there's a problem with the teleprompter? What if it's not in the right space? What if someone stands up? What if someone boos me? All of those factors are not in virtual reality. We have to learn how to manage in the real world and deal with those scenarios. Definitely, there's a lot of value by using simulations in virtual reality to help us practice and understand the task we're looking to do. It definitely gives us confidence, and it definitely takes away some uncertainty in those situations. The question is this: How can we prepare for the spontaneity which is the human experience?

The Haves and the Have-Nots

We have the digital natives, and we have the digital immigrants, or the individuals who were born before the Internet. They were born before 1980 and have a life before the smartphone and the Internet. The digital natives are those who were born into the age of the Internet. All they know is the Internet, smartphones, and computers. This is their world. In the world of digital natives, we have something called digital diffusion in digital inclusion. We have those who are digital natives and don't have access to adequate technology.

There's a divide between those who have access to the latest and greatest technology and those who do not. If we look at it on a global scale, it ties into virtual reality in the sense that there are those who can afford to have virtual reality technology or at least access to the technology, perhaps outside of the home, at school, or at any type of clubs or educational environments that are outside of their home that give them access. They can afford access to the technology directly or indirectly.

Then there are those who have no access to the technology at home or outside of the home, either directly or indirectly. There's a disparity among digital natives when it comes to virtual reality. There are those who have and those who have not. It's important to understand that as the technology improves and moves forward, those who have access to the technology, whether directly or indirectly, have an advantage over those who do not have access to the technology. Because they have exposure, they're able to navigate it more quickly, and when it is in terms of the marketplace, they have experience, especially when we look at the marketplace in the United States. Employers look for individuals with experience. They want someone to start a job and be able to literally begin on day one with little or no learning curve.

As virtual reality increases in use in the marketplace, this disparity will impact job positions for those who do not have either direct or indirect exposure to virtual reality technology. This is something that needs to be addressed, and it can carry over not only to virtual reality technology but to all technology. If we have students who have access at home to Microsoft Office and are able to use Microsoft Office Suite and have experience with it, whether at home or at school, they have a leg up on students or individuals who do not have access to it at home and do not have access to it outside the home to give them that experience. Now they're in the workplace, innocent and challenged because they have to learn how to use the technology when there are others who can sit right down and get to work. Many times, those who are exposed to these technologies and have experience may be even ahead of their employer, which gives them extra value in the workplace and the marketplace because of their experiences with the technology. They can bring new ideas, and they're more efficient.

These are the things to remember when we talk about the haves and the have-nots. It is a strong point because what begins to happen is we not only have a community but also a society where there is not only an economic class system but also a technology class system. We have to forward think, especially with technology evolving, changing, and developing as fast as it is today.

Exposure versus No Exposure

It is important to consider that exposure to technology is just as important as actually being fully immersed in the technology. It's important for students to at least have the awareness that certain technologies, especially virtual reality, exists. We already agreed that virtual reality is a displacer, and it is disrupting as we speak. So this is not a future endeavor or technology that will impact us in three years or five years. This is something that is impacting us currently. Students need to have some type of exposure or use case with virtual reality. I tend to think of it as when we talk about minimum viable product (MVP), which can be something as simple as a PowerPoint presentation that outlines the whole process or the way the product will be implemented. Or an MVP could be a video explaining how the product could be used without having the actual or working prototype of the product. The reason for this is that it is cost-effective and gives the potential customer or prospect the ability to see what the product is and give feedback.

The important thing with virtual reality at this point is that there are costs involved. It does prohibit access; however, we are able to allow students to experience even at the slightest or lowest level the technology that allows them not to give feedback but to have that thought pattern and have access to an understanding of it so as we fast forward, they're not starting at zero or with a blank slate.

Exposure is important because as technology continues to develop faster and faster, it becomes challenging to actually learn technology and master it. What I mean by this is that technology keeps advancing, so by the time we learn one form of technology, a new one has already come out, and it becomes a point of learning skills. We talked about this earlier, how you can learn a technology, and then there's Tableau, there's Abacus, there's other software out there that can do data analysis. Was the key skill that you need to know to utilize all these software programs exposure to the software, which is the gray area? Or is it learning the skill of programming? When we talk about virtual reality, it is gray. Learning how to program or use software and hardware is amazing, but the key concepts are the skills, the use cases, the understanding, and the depth of the technology. They are really what are going to push forward and allow our students and users to be able to move forward.

Exposure allows us to see what's possible. Exposure gives us a glimpse into the future. Exposure gives a foundation for those to be able to move forward, so it is imperative that we have exposure to our technology, our virtual reality technology, and our understanding. Then we can start getting the skills we need to use them and take the technology to the next level in the future.

The more exposure, the better. The more exposure students have to new ideas and new virtual reality technology, the better they are, because that exposure gives them a foundation to work with. Think about Bill Gates. He was in high school with exposure to a computer and mainframe computer in the 1960s, which allowed him to program. Ultimately, the students at his school started a programming club that sparked them to move further ahead in programming and become experts in it. So exposure is very important for the development of virtual reality. The more individuals, especially students, who are exposed to it, the more potential ideas are able to take root.

Technology versus Teacher

We have teachers who are digital immigrants, and we have students who are digital natives. That is true from kindergarten through 12th grade and all the way through higher education. The problem is that we have educators who are still trying to teach students in their own traditional ways. Then we have students who are not engaged in the traditional ways because they are digital natives.

As educators, we must learn how to reach our students. We must learn to understand how they think and how they work and how they learn, which is completely different than what we've been used to. Virtual reality is a new technology to the digital immigrant and even to many digital natives. As educators, we need to know more about the technology than our students in order to teach them, expose them, and allow them to use the technology. This is our challenge as educators. We must be ahead of the curve, not at the curve because technology is constantly changing. We have to be at the fore-front or at least in the door when it comes to new technology such as virtual reality. We need to know what the newest trends are in an attempt to anticipate what's coming next so we can expose our students to the new technology. That doesn't mean we will have 100%. We may not have the new technology, the headsets, and the software for all our students, but we are able to expose them to the ideas, the process, and the content. I look at it as if we were showing students an MVP (minimum viable product), and that could be anything from a full-blown model to a PowerPoint with slides showing the process we're looking at. This engages them and allows them to see what's coming next.

As educators, we need to take responsibility for what we're doing. As educators, we need to make sure we maintain our current knowledge of technology either through professional development, which is something that happens to us and is usually paid for by our institutions, or through professional learning that should be ongoing, development that is initiated by us as educators. Professional learning can be online, at conferences, or face-to-face. It can even be through podcasts or YouTube videos as long as we're getting information, particularly with virtual reality. As the technology continues to develop, we need to stay focused on being up-to-date on new hardware as well as new software and applications that can be used in the classroom in order to bridge the gap for many of our students and help them learn and have access to the technology whether it's from Oculus Rift or using something on their smartphones through an application. We should have knowledge of that and be utilizing it.

As educators, we need to cultivate a spirit and mentality of innovation. We need to have the mindset that we have to be ahead of our students. We need to know more than our students in order to teach them. Technology is continuously changing, faster than we can keep up with, so it may be stressful and cause anxiety, but we have to realize that in order to serve our students and society, we have to cultivate a culture of innovation.

We have to be able to think forward and use the resources that are available to us. Many times we become complacent and used to the way things are. We may not see two steps ahead of us or believe that certain things may move forward and become commonplace. Again I'll use the example of Uber. Since it was founded in 2009, the way we travel in cities has totally changed. The word Uber is now a verb for ridesharing. At one point, it was not culturally acceptable to get into a stranger's car, and taxis dominated the industry. As we see it now, taxis are no longer a dominant force, especially for transportation in major cities. It's Uber everywhere you go on the planet, and people are commonly using rideshare. It is something that we have to think about. Technology can move more quickly than we can possibly imagine and can displace the prevalent technology in a moment's notice. How we prepare students for that type of world only gets faster.

When we think about virtual reality and the advent of the technology, let's look back to the advent of the movable type printing press in the 1500s. That new invention opened up the way for many books to be printed and for more people to read and have access to the printed word. As we see the price of virtual reality technology decrease and there's more access, there will be more hardware and an increase in software because we'll have more individuals who are able to use the technology. There will be an increase, and as the increase moves forward, there will also be more innovation, so we need to think about how things moved in the past, which will impact the future. We also need to be forward thinking and not be so afraid or think so narrowmindedly that the technology will not increase and more and more people will use it. There is a steady increase in the use of virtual reality hardware, which means there will be an increase in the software, and that means more and more people will be creating applications, whether it's games or other uses. Use cases will develop for the technology. These are the things we need to keep in mind as we move forward in time and as virtual reality continues to spread and move forward.

We also need to look for training. We need to train our students. With that in mind, know that the world we live in today is going to be totally different than the world is tomorrow. Always go back to the Uber and Lyft examples, that about 10 years ago they were not culturally accepted and not widely accepted modes of transportation. Almost anywhere you go in the United States, let alone the world, you can take an Uber or Lyft. It's a common mode of transportation for most people young and old. When we think about this shift, we can contend with virtual reality right now. We may not see much of it as far as the mainstream public, but this can change very quickly, and we need to prepare ourselves and our students for this new world.

We must be forward thinking, not only for ourselves but for our students and future generations. Let's push the envelope and stay ahead of the curve. I know I may sound like I'm beating a dead horse and it may seem like common sense, but in practice, it's very hard to do. It's very hard for us to accept change. Most people do not like change and like to stay the same.

Unfortunately, the world we live in is changing technologies, and virtual reality is a new platform, a new area not only in productivity but also in education. It's changing the way we communicate and interact with the world, and it will ultimately bring us closer. I look at the pioneers of the Internet age and how they moved forward with PCs. We need to be aware of the expansion of virtual reality. Just as the PC enabled the age of the Internet, virtual reality, as the hardware goes down in price, will enable the age of virtual reality to increase and make virtual reality and augmented reality move forward. We already see how augmented reality is becoming very commonplace with access to hardware such as the smartphone.

About the Author

Mfon Akpan is a professor of business at the Savannah College of Art and Design (SCAD). He continues to research new techniques and educational methods to offer students and clients current and practical engagement tools. He is an expert in implementing virtual reality technology into college courses.

Visit Mfon at *MfonAkpan.com*
Watch Mfon's TEDx on Virtual Reality in the Classroom at
MfonAkpan.com/speaking-workshops